Pre Algebra Workbook 6th Grade:

Exponents

(Baby Professor Learning Books)

Speedy Publishing LLC
40 E. Main St. #1156
Newark, DE 19711
www.speedypublishing.com

Exponents

Level 1

Solve.

1. 7^1

2. 7^2

3. 3^2

4. 6^1

5. 9^2

6. 10^9

7. 8^1

8. 100^3

9. 1^{22}

10. 1^{7}

11. 10^{7}

12. 4^{1}

13. 8^{2}

14. 1^{61}

15. 0^{83}

16. 0^{96}

17. 0^{26}

18. 3^{1}

19. 7^1

20. 1^{32}

21. 1^{81}

22. 5^1

23. 9^1

24. 10^7

25. 10^5

26. 8^2

27. 0^9

28. 100^5

29. 0^{21}

30. 2^1

31. 1^{78}

32. 3^2

33. 6^1

34. 7^2

35. 5^3

36. 4^2

37. 6^2

38. 2^3

Exponents

Level 2

Solve.

1. 0^{25}

2. 6^{1}

3. 7^{2}

4. 1^{48}

5. 94^{1}

6. 2^{10}

7. 54^{2}

8. 9^{2}

9. 2^9

10. 95^1

11. 93^2

12. 100^5

13. 1^{67}

14. 44^2

15. 82^1

16. 10^6

17. 6^1

18. 3^1

19. 4^1

20. 1^{45}

21. 26^1

22. 66^1

23. 6^2

24. 100^3

25. 7^2

26. 9^1

27. 77^2

28. 9^2

29. 93^2

30. 85^1

31. 14^2

32. 48^1

33. 2^9

34. 3^2

35. 48^2

36. 1000^2

37. 3^2

38. 88^2

39. 8^2

40. 5^2

41. 56^2

42. 84^1

43. 85^2

44. 10^{10}

45. 2^4

46. 10^{11}

47. 1^{83}

48. 1^4

49. 8^2

50. 15^2

51. 69^2

52. 5^1

53. 29^1

54. 5^4

55. 30^2

56. 49^2

57. 7^2

58. 50^1

Exponents

Level 3

Solve.

1. 7^0

2. 4^{-3}

3. 1^{21}

4. 10^0

5. 100^0

6. 0^{15}

7. 10^{-7}

8. 4^2

9. 7^{-2}

10. 10^{-3}

11. 0^{52}

12. 100^{6}

13. 10^{7}

14. 100^{-2}

15. 5^{-3}

16. 2^{5}

17. 7^{-2}

18. 9^{0}

19. 1^{74}

24. 0^{0}

20. 1^{30}

25. 3^{-2}

21. 100^{4}

26. 8^{-1}

22. 2^{-4}

27. 5^{0}

23. 10^{9}

28. 9^{0}

29. 4^{-2}

30. 2^0

31. 4^{-1}

32. 8^0

33. 4^0

34. 0^0

35. 3^0

36. 7^0

37. 1^{75}

38. 8^{-1}

39. 4^2

40. 1^{49}

41. 6^1

42. 2^6

43. 2^3

44. 10^{-2}

45. 0^1

46. 10^3

47. 2^0

48. 4^{-3}

49. 100^3

54. 5^{-1}

50. 1^0

55. 4^{-1}

51. 2^0

56. 10^7

52. 1^{81}

57. 7^0

53. 9^1

58. 4^0

Exponents

in equations level 1

Solve.

1. $0^{36} \times 1^{49}$

2. $2^{6} + 7^{2}$

3. $0^{27} - 5^{3}$

4. $3^{1} + 2^{3}$

5. $9^{2} / 2^{4}$

6. $0^{59} - 3^{2}$

7. $1^{90} - 4^{2}$

8. $7^{1} + 0^{75}$

9. $1^{100} + 8^2$

14. $100^5 / 1^{32}$

10. $2^5 + 3^3$

15. $6^2 - 10^7$

11. $8^1 - 4^3$

16. $100^6 - 5^2$

12. $100^1 - 1^{92}$

17. $5^1 - 1^{63}$

13. $0^9 - 8^1$

18. $100^4 \times 0^{33}$

19. $6^1 + 10^6$

20. $1^{25} - 10^3$

21. $2^5 - 9^1$

22. $8^2 - 2^4$

23. $5^3 / 4^1$

24. $1^{35} \times 1^{46}$

25. $5^2 + 0^{77}$

26. $7^2 - 6^1$

27. $100^3 - 7^1$

28. $0^{75} / 1^{38}$

29. $10^2 + 10^4$

30. $1^{57} + 10^7$

31. $9^2 / 3^3$

32. $100^2 \times 2^6$

33. $7^2 - 4^1$

34. $5^1 - 100^4$

35. $1^{40} + 1^{27}$

36. $5^3 + 1^{18}$

37. $1^{97} + 3^3$

38. $8^1 + 4^3$

39. $1^{40} - 5^{2}$

40. $3^{2} + 7^{1}$

41. $9^{2} - 4^{1}$

42. $0^{33} / 4^{2}$

43. $3^{1} \times 10^{3}$

44. $2^{5} - 100^{2}$

45. $4^{3} / 2^{1}$

46. $2^{4} \times 2^{6}$

47. $1^{2} + 0^{71}$

48. $0^{98} \times 100^{3}$

49. $3^3 \ / \ 1^{45}$

50. $8^1 + 1^{75}$

51. $9^1 \times 1^{40}$

52. $5^3 \times 0^{28}$

53. $1^{39} - 4^3$

54. $6^2 \ / \ 2^5$

55. $7^2 \times 10^{10}$

56. $1^{86} \times 10^4$

57. $2^6 + 3^2$

58. $2^1 + 10^8$

Exponents

in equations level 2

Solve.

1. $(-1)^8 + 4^2$

2. $(-1)^{90} + 1^{99}$

3. $(-7)^2 \times 10^5$

4. $3^3 / 7^1$

5. $0^{15} - 0^9$

6. $(-100)^1 + 0^{98}$

7. $(-1)^{19} + 9^2$

8. $8^1 - 5^3$

9. $(-2)^6 \,/\, (-3)^2$

10. $(-2)^3 - (-2)^1$

11. $(-5)^1 - 0^{54}$

12. $(-2)^5 + (-8)^2$

13. $2^3 \times (-2)^6$

14. $(-10)^1 - 0^{96}$

15. $(-2)^1 - (-5)^2$

16. $1^{71} + (-1)^{90}$

17. $1^{15} - (-6)^2$

18. $100^1 + 0^{49}$

19. $2^4 \times 0^{24}$

20. $4^2 + 5^1$

21. $1^{49} + (-7)^2$

22. $4^3 - (-9)^2$

23. $1^{28} / (-10)^3$

24. $4^1 / (-1)^{38}$

25. $(-2)^3 / (-9)^2$

26. $8^2 / (-3)^3$

27. $3^2 \times (-6)^1$

28. $1^2 - (-1)^{50}$

29. $7^2 \ / \ (-8)^1$

30. $1^{38} + (-5)^1$

31. $10^7 - (-1)^{19}$

32. $1^{56} \times (-1)^{30}$

33. $100^5 \times (-5)^2$

34. $(-7)^1 \times 100^6$

35. $0^{70} + (-1)^{17}$

36. $10^1 \times 0^{52}$

37. $(-7)^1 - 10^4$

38. $(-9)^2 \ / \ (-100)^5$

39. $(-8)^2 - 100^3$

40. $0^{37} - 0^1$

41. $7^2 / 2^1$

42. $(-3)^2 \times (-1)^{91}$

43. $2^5 / (-1)^{71}$

44. $(-4)^1 - (-4)^3$

45. $(-2)^6 - 3^1$

46. $(-1)^{42} - 3^3$

47. $10^5 + (-100)^4$

48. $(-1)^{18} - (-8)^1$

Answers

1. 7
2. 49
3. 9
4. 6
5. 81
6. 1000000000
7. 8
8. 1000000
9. 1
10. 1
11. 10000000
12. 4
13. 64
14. 1
15. 0
16. 0
17. 0
18. 3
19. 7
20. 1
21. 1
22. 5
23. 9
24. 10000000
25. 100000
26. 64
27. 0
28. 10000000000
29. 0
30. 2
31. 1
32. 9
33. 6
34. 49
35. 125
36. 16
37. 36
38. 8

1. 0
2. 6
3. 49
4. 1
5. 94
6. 1024
7. 2916
8. 81
9. 512
10. 95
11. 8649
12. 10000000000
13. 1
14. 1936
15. 82
16. 1000000
17. 6
18. 3
19. 4
20. 1
21. 26
22. 66
23. 36
24. 1000000
25. 49
26. 9
27. 5929
28. 81
29. 8649
30. 85
31. 196
32. 48
33. 512
34. 9
35. 2304
36. 1000000
37. 9
38. 7744
39. 64
40. 25
41. 3136
42. 84
43. 7225
44. 10000000000
45. 16
46. 100000000000
47. 1
48. 1
49. 64
50. 225
51. 4761
52. 5
53. 29
54. 625
55. 900
56. 2401
57. 49
58. 50

1. 1
2. 1/64
3. 1
4. 1
5. 1
6. 0
7. 1/10000000
8. 16
9. 1/49
10. 1/1000
11. 0
12. 1000000000000
13. 10000000
14. 1/10000
15. 1/125
16. 32
17. 1/49
18. 1
19. 1
20. 1
21. 100000000
22. 1/16
23. 1000000000
24. 1

25. 1/9
26. 1/8
27. 1
28. 1
29. 1/16
30. 1
31. 1/4
32. 1
33. 1
34. 1
35. 1
36. 1
37. 1
38. 1/8
39. 16
40. 1
41. 6
42. 64
43. 8
44. 1/100
45. 0
46. 1000
47. 1
48. 1/64
49. 1000000
50. 1
51. 1
52. 1
53. 9
54. 1/5
55. 1/4
56. 10000000
57. 1
58. 1

1. 0
2. 113
3. -125
4. 11
5. 5
6. -9
7. -15
8. 7
9. 65
10. 59
11. -56
12. 99
13. -8
14. 10000000000
15. -9999964
16. 999999999975
17. 4
18. 0
19. 1000006
20. -999
21. 23
22. 48
23. 31
24. 1
25. 25
26. 43
27. 999993
28. 0
29. 10100
30. 10000001
31. 3
32. 640000
33. 45
34. -99999995
35. 2
36. 126
37. 28
38. 72
39. -24
40. 16
41. 77
42. 0
43. 3000
44. -9968
45. 32
46. 1024
47. 1
48. 0
49. 27
50. 9
51. 9
52. 0
53. -63
54. 1
55. 490000000000
56. 10000
57. 73
58. 100000002

1. 17
2. 2
3. 4900000
4. 3 6/7
5. 0
6. -100
7. 80
8. -117
9. 7 1/9
10. -6
11. -5
12. 32
13. 512
14. -10
15. -27
16. 2
17. -35
18. 100
19. 0
20. 21
21. 50
22. -17
23. - 1/1000
24. 4
25. - 8/81
26. - 2 10/27
27. -54
28. 0
29. - 6 1/8
30. -4
31. 10000001
32. 1
33. 250000000000
34. -7000000000000
35. -1
36. 0
37. -10007
38. - 81/1000000000
39. -999936
40. 0
41. 24 1/2
42. -9
43. -32
44. 60
45. 61
46. -26
47. 100100000
48. 9

CPSIA information can be obtained
at www.ICGtesting.com
Printed in the USA
BVHW091426171222
654335BV00008B/515

9 781682 800485